Can you count and tell shapes?

By Caroline Sogomo RN-BC

All Rights Reserved. ©Copyright Sogomo Global Inc. 2015

Disclaimer
No part of this book may be reproduced or transmitted in any form or by any means, mechanical or electronic, including photocopying or recording, or by an information storage and retrieval system, or transmitted by email without permission in writing from the publisher.
This book is for entertainment purposes only.
The views expressed are those of the author alone

Number Recognition

ONE
MOJA

Number Recognition

TWO

MBILI

Number Recognition

THREE
TATU

Number Recognition

FOUR

NNE

Number Recognition

FIVE
TANO

Number Recognition

SIX

SITA

Number Recognition

SEVEN
SABA

Number Recognition

EIGHT

NANE

Number Recognition

NINE

TISA

Number Recognition

TEN
KUMI

Number Recognition

ELEVEN

KUMI NA MOJA

Number Recognition

TWELVE

KUMI NA MBILI

Number Recognition

THIRTEEN

KUMI NA TATU

Number Recognition

14

FOURTEEN

KUMI NA NNE

Number Recognition

FIFTEEN

KUMI NA TANO

Number Recognition

SIXTEEN

KUMI NA SITA

Number Recognition

SEVENTEEN
KUMI NA SABA

Number Recognition

18

EIGHTEEN

KUMI NA NANE

Number Recognition

19

NINETEEN

KUMI NA TISA

Shape Recognition

CIRCLE

DUARA

BALL

MPIRA

Shape Recognition

DIAMOND
MSAMBAMBA

KITE
KISHADA

Shape Recognition

HEART
MOYO

HEART
MOYO

Shape Recognition

OVAL
DUARADUFU

EGG
YAI

Shape Recognition

RECTANGLE
MSTATILI

CELL PHONE
SIMU

Shape Recognition

SQUARE

MRABA

DICE

KETE

Shape Recognition

STAR
NYOTA

STAR
NYOTA

Shape Recognition

TRIANGLE
PEMBETATU △

ICE CREAM
AISKRIMU

THE END !

MWISHO !

www.ingramcontent.com/pod-product-compliance
Lightning Source LLC
Chambersburg PA
CBHW041119180526
45172CB00001B/328